WHAT!
MEMORY?

SOME THINGS NEVER FADE

I0504122

CHANDHANA
SATHISHKUMAR

ISBN 978-1-68523-953-4

This book is dedicated to everyone with a red line under their name on Google Docs

Contents

COPYRIGHTS

FOREWORD

"If our brains were simple enough to be understood, we
wouldn't be smart enough to understand them."
— David Eagleman

CONTENT

I

The Basics

Memories are stored as information in our neurons. Well, not in neurons themselves, but rather in the network made of millions of neurons. The basis of memory storage is how two neurons connect. Learning is essential to coordinate behavior. It means an isolated neuron is of no use.

But what's the story behind it? There's a whole world of microscopic chemical changes to unravel to understand the functioning of memory.

When a specific pattern of neurons fires, there is a specific output, rather it is a movement, memory, or computation. It all begins in encoding. What's encoding you may wonder. In terms of computers it's when you change a file, say for example a wav file, to an mp3 file because your computer isn't compatible with wav formats. Similarly, in terms of memory, things you see, touch, feel or experience is converted into information that means something to you so you can remember it. This is done through semantic processing wherein the perceptive features of memory travel to the hippocampus separately.

Then they are all integrated as one complete experience. It would be quite unrealistic for us to remember every single moment of our lives(it is possible though). So as soon as we experience something the information gets encoded in our "short term memory".

Once the memories enter this area they have two options- to either be transferred to our long-term memory or they can be forgotten. Your frontal cortex and hippocampus decide whether the experience is worth remembering. Memory is stored when the hippocampus takes short-term memory and transforms it into long-term memory through alterations in neural networks, proteins, gene expression, etc.

Long-term memory has an unlimited capacity to retain information/experiences for large amounts of time. Going over something, again and again, will strengthen the network so that memory becomes stronger. That's why reviewing things over and over again will let you remember things better.

❦

There are two different types of memories here:

Unconscious and conscious memory. As the name suggests, Unconscious memory is acquired and used unconsciously. They are slower to acquire but harder to lose, thus shaping the way we think, the way we act, and ultimately our entire personality. You can conclusively say that it is the primary source of human behavior. For Eg, bathroom singing your favorite song, brushing your teeth, typing, the way you talk to people.

On the other hand, Conscious memory usually includes facts, names, and dates and can be lost just as quickly as it is obtained.

If encoding and storage are the main steps for withholding memories within the brain, what happens when we need to recall one?

This stage is called the retrieval stage and it refers to the ability to recall or retrieve information that has been stored beforehand. All you need to do is contact the unconscious level of LTM (Long Term Memory) storage and it will then be transferred to the conscious level (or your working memory). It's safe to arrive at a conclusion that Retrieval and rehearsal of data go hand in hand.

Human memory is far from perfect. Memory is considered an active reconstructive process because every time we recall a memory, there's usually something we rebuild. Be it a blurred detail lost to time or a complete figment of our imagination.

II

The Biology of
Memory

The hippocampus is the part of the brain associated with recalling memories. Animals have one on either side of the brain. Located in the medial temporal lobe (structures that are key in memory recollection), the hippocampus is also key in spatial navigation and regulating emotions.

Just like the hippocampus, the amygdala is part of the limbic system and is located at the end of the hippocampus. The amygdala works in tandem with the hippocampus in memorizing emotions, especially fear. That's why you feel different emotions whenever you recall different memories to the surface.

Located in the lower area of the brain, the cerebellum is responsible for the balance and coordination of the body muscles. It allows one to perform everyday functions and helps one stay upright. It is key in the processing of procedural memory (how to perform certain skills).

To help know what Acetylcholine is, here is an analogy. Consider memories of passengers and the previous three brain components stations, acetylcholine is the train that allows memories to reach these stations. Specifically, it is a neurotransmitter, one of many chemicals that neurons use to communicate with one another. So it is this fuel that allows us to store and recall memories. People with a lack of the chemical (such as those with Alzheimer's) are unable to recall or store memories properly.

A concept in psychology, localization of function basically refers to the idea that different parts of the brain are responsible for certain functions or specific behaviors. So damage to the memory storage parts (such as the hippocampus and amygdala) might render one incapable of memorizing things, even though their perception of such events is still perfect.

Neuroplasticity is a much broader concept, It refers to the process in which your brain's pathways and synapses become altered as a result of changes in environmental, behavioral, and neural conditions. Neuroplasticity takes place throughout your entire lifetime and mainly aims to wire the programming of your brain for maximum efficiency. Neuroplasticity also works with synaptic pruning to ensure your brain only has what it needs to be efficient. Synaptic pruning essentially cuts off any unnecessary information or skills in favor of developing those that you need.

Neurons are the cells that carry electrical impulses to and from larger modules in your brain. The pathways that connect neurons are known as synapses, where they can pass the electrical or chemical signals to other neurons.

The Nervous system is essentially your lifeline. Without your nervous system, your body would be a corpse, unable to move or do anything. The nervous system is a complex network of cells and nerves that carry messages to and from the brain and spinal cord to various parts of the body.

Within the more scientific terminology, the nervous system is made of the Central nervous system (CNS) and the Peripheral nervous system. The CNS is made up of the brain and spinal cord while the PNS is made up of Somatic and Autonomic nervous systems.

III

Brain Breakdown

Neurons are the basic units here; they receive and transmit the signals from the brain and spinal cord to where they need to go. The central nervous system essentially controls the activities of the body. The somatic nervous system picks up sensory inputs from limbs and other distant organs separate from the brain and carries them to the nervous system.

For example, if your hand were to cut itself on a sharp object, the somatic nervous system would carry information about the cut to the brain, which would then send the signal back to the affected hand telling it to reel back immediately. The Autonomic nervous system, however, is almost invisible to us. That's because the autonomic nervous system controls the nerves and functions of the inner organs, on which we have no conscious control (these include breathing, digestion, heartbeats, etc).

Neurons are made up of three main components:

1. Soma (Cell body) — Contains genetic information, produces peptide neurotransmitters, and creates vesicles that carry neurotransmitters
2. Dendrite — The receiving end of a neuron that is typically characterized by many short branches
3. Axon — The transmitting end of a neuron that is typically characterized by a long branch coming from the cell body

❧

Neurotransmitters are instrumental in the process of dendrites receiving and axons relaying information. Between neurons, there are junctions that enable space between the two cells. When a signal has to be passed onto the next neuron(s), neurotransmitters are released from the presynaptic axon, cross the synaptic cleft, and bind to receptors on the postsynaptic dendrite.

There are two main types of neurotransmitters:

1. Inhibitory — Hyperpolarize (decrease) the membrane potential of the postsynaptic neuron which prevents it from firing
2. Excitatory — Depolarize (increase) the membrane potential of the postsynaptic neuron which may lead to the neuron firing

IV

Schemas

J. Piaget, a Swiss psychologist known for his work on child development, cognitive development, and genetic epistemology introduced to us an important theory commonly called the schema theory or a schema. Schemata refers to the abstract concepts of webbing complex relationships with one another. They include thoughts, perspectives, facts, stereotypes, and archetypes.

For example, picture a person. You immediately think of a human being. Two arms, legs, a head, and so on. Now, what if I describe our imaginary human even more? What if I said our human is female. Say about 17 years old. You can see a clear shift from your initial thoughts or scheme of a random person, possibly even someone you know to something more specifically defined. How about Tanned skin and Blue hair? Indian? Hispanic? Maybe multi-ethnic?

Have you noticed when you think about a certain idea, your mind immediately brings up related information like the recommended movies on Netflix? Just like our dear flix has its Recommendation Algorithm our brain too has a system that links data to related facts. And our brain's system of recommendation is waaaayyy more intricate.

These examples are very broad, non-specific. Our schemas can get insanely complex as we learn and experience more things by the minute. The more we learn and specialize in a certain field the easier it is for us to link our newly learned information to the pre-existing data, and thus attach to it. They play a vital role in how memories are encoded and retrieved and also how the details are deciphered.

In simple terms, Schemas pretty much refer to the information we already know. It might be the basics of a certain topic, the lyrics of a song, a book you read, or the way to school or work. What we know might be incorrect or insufficient since schemas are made of what a person knows and is unique in every possible way. Schemas, just like the human memory, are not completely reliable. When we learn concepts in-depth or correct our mistakes we overwrite the pre-existing information as a correction or simply add on to it leading us to build a framework for future perceiving and understanding of information. This way, the Schema web grows infinitely and hence can be adjusted or refined throughout the span of our lives.

V

Types of Memory

Long-term memory is divided into two sections: explicit memory and implicit memory. Explicit or declarative memories are ones that you actively seek out in order to recollect and share. Typically, this comprises long-ago events or knowledge that relates to the current issue. It's a form of long-term memory that necessitates conscious thought. When most people think of a memory, this is what comes to mind.

Implicit memory, on the other hand, is the polar opposite. It's also known as unconscious memory since it relies on prior experiences or knowledge to recall things without you having to think about them. It's a type of long-term memory that doesn't need conscious cognition. It enables you to perform tasks by rote.

Declarative memory is similar to explicit memory in that it deals with consciously remembered experiences and information, such as facts, data, and events. Procedural memory, on the other hand, is classified as implicit memory since it is used subconsciously anytime you execute or acquire a skill.

One of the two forms of declarative memory is semantic memory (wow, this is getting confusing). Semantic memory refers to the general world information we've collected throughout the course of our life, whether or not we've been actively involved in it (i.e naming capitals of all countries). The opposite of episodic memory is autobiographical memory, which is listed here (people, events, times, emotions, etc.).

Eidetic memory refers to the ability of the brain to recall vivid pictures that are nearly similar to a genuine thing. This memory kind is exceedingly uncommon, and it does not occur in adulthood.

The medical word for a photographic memory is hyperthymesia. People with hyperthymesia (yes, they exist) have an abnormally high level of precision and accuracy while recalling their life events.

Reconstructive memory works by piecing together previously remembered facts and information to create a picture of what happened. This happens when you don't have a cohesive or clear recollection to help you. When a detective utilizes information from a witness to recreate the events at a crime scene, this is an example.

Have you ever wondered why your fingers are so proficient at typing on the keyboard? Or why your legs are so tuned in to a certain swimming stroke? You can thank muscle memory for that. This type of procedural memory involves repeatedly committing a specific motor task to memory.

Two or more people in a social group have shared knowledge and information. Because more people may consolidate existing knowledge or add on new pieces, groups are more dependable than individuals at retaining and retrieving memories. As a result, these experiences are frequently more detailed in terms of remembrance. For example, the United States commemorates 9/11, the United Kingdom commemorates the British Empire, and Jews commemorate the Holocaust.

Genetic memories are those that exist in the absence of sensory awareness and are found within a human at birth. This is essentially the concept that a group's shared experiences are encoded in their genetic code and handed down to their offspring's genomes over extended periods of time. Some believe that language belongs in this category since it is a trait present in the neurological system at birth, but the perception of a language's phonemes (distinct units of sound) is integrated much earlier. Traumas, phobias, and neuropsychiatric illnesses are the most frequent genetic memories. In a research study, scientists discovered that mice educated to dread a certain fragrance passed that fear on to their offspring, despite the fact that the mice had never encountered the smell or been trained to fear it.

VI

Flashbulb Memory

What are flashbulb memories? How are they formed, and what differentiates them from other memories? Do you think it strange that when anything unpleasant or emotionally stimulating occurs, you can retain clear memories of your whole doings and locations? That's a recollection that came back to me like a flashbulb. Flashbulb memories are extremely precise autobiographical recollections of experiences that were startling or even stimulating, similar to how a camera flash clearly captures a single instant. When someone claims they recall a horrific incident such as an assassination or a natural disaster, they usually mean they remember learning about it after it occurred.

Flashbulb recollections are memories of discovering something so surprising or significant that it leaves you with a strong and accurate recall of learning about the event, but not necessarily the event itself. Flashbulb memories are distinct from other types of memories in that they rely on personal significance, repercussions, emotion, and surprise. All of these elements are more prevalent in flashbulb recollections than in traditional autobiographical ones. Place, continuing action, informant, own effect, another effect, and aftermath are the six distinguishing aspects of flashbulb memories. The first is where they were when they learned of the incident, the second is what they were doing at the time, and the third is who delivered the word to them. The term "your impact" relates to how you reacted to the news, and "another effect" refers to how the rest of society reacted. The third stage is the aftermath, in which your brain remembers what occurred to you and others as a result of the news.

Why do we remember some memories more vividly than others? When we feel something is important to us, the brain makes sure to keep the information safe, and because we have strong emotions attached to those memories we remember them more vividly than others. The level of surprise, consequentiality, and even emotional arousal that flashbulb memories produce is the fundamental reason why they exist in the first place. A horrific experience is considerably more likely to stick with you than a mundane news item. It doesn't have to have anything to do with you personally, but most occurrences are.

Why are some cultures more likely to forget events as a whole? Some cultures are more likely to forget an event as a whole as the event might not to talked about to future generations. For example, many Americans don't know the reason behind the celebration of Krishna Jayanti or Halloween. This may be so because it isn't significant or has not impacted them the way it has in other cultures.

If such cultures believe the events to be too painful or desire to forget them, they will, in fact, purposefully forget them. For example, because the Korean War was such a debacle for the US government and military, the people were not informed about it. Because of widespread rejection of any remembrance of the Korean War, a whole generation was raised without hearing about it. It is still known as the "forgotten conflict."

Overt rehearsal is the word used when someone tries to rehearse memories or abilities in front of others, making it obvious that they are remembering something. Covert rehearsal is the polar opposite; it generally entails silently recalling a memory such that no one else is aware you're doing it. Which one you should utilize depends on the scenario and your recall. If you're attempting to memorize the words of a song, for example, it's preferable to sing them out loud . If it's just a basic fact or information recital, though, a quiet and hidden rehearsal is generally preferable.

Individualistic and collectivistic are words that refer to cultural styles rather than memories. Individualistic cultures (such as those seen in Western countries like America) value individual expression and innovation. Collectivistic cultures (such as those seen in China or Japan) place a higher value on family, kinship, and community above individual thought.

Emotional arousal, in scientific terms, is a heightened state of psychological activity, comparable to being woken. It fundamentally controls how we react to ordinary situations and how we experience them. It also influences how we categorize things in relation to the feelings we experience as a result of them.

Returning to the notion of flashbulb memory, Neisser and Harsch sought to determine how dependable this form of memory was (since some were claiming it was far higher in terms of accuracy when recalling it). 106 students were handed a questionnaire less than 24 hours after the Challenger space shuttle tragedy, which contained seven questions about where they were, what they were doing, and so on. After two and a half years, 44 of the original students returned to complete the survey, but this time they were also asked to assess their confidence in their memory's accuracy on a scale of 1 to 15. Participants were also asked if they had previously completed a questionnaire on the issue.

The two then did a semi-structured interview to see if they remembered what they'd written earlier. Finally, their replies from two years ago were displayed to them. The findings were startling, demonstrating how incorrect even flashbulb recollections may become with time. Only a quarter of the participants remembered taking the original survey. The correctness of memory of the seven questions had a mean score of 2.95 out of 7. Eleven individuals who received a score of 0 and 22 had a score of 2 or less. For the questions, the average degree of confidence in correctness was 4.17. As a result, Neisser and Harsch came to the conclusion that flashbulb memories were not as accurate or long-lasting as previous theories suggested.

Brown & Kulik set out the idea that Neisser and Harsch would question a decade later, a decade before Neisser and Harsch conducted their experiment. A questionnaire was provided to 80 participants in which they were asked to recollect situations in which they had learned of upsetting incidents. The participants, as predicted, had clear memories of where they were, what they were doing, and how they felt when they learned of a horrific occurrence. Personal tragedies, such as the death of a loved one, have also been linked to flashbulb memories.

Brown and Kulik deduced from this that flashbulb memories were more likely to occur during unexpected and intimate situations and that these recollections were triggered by psychological-emotional arousal . Collectivistic societies have lower levels of emotions, mental ruminations , and social sharing of emotions, according to Wang and Aydin. They also discovered that perhaps more crucially, people in these cultures did not recall as many details about public events utilizing flashbulb memory as people in individualistic civilizations.

VII
Memory Aids

Using elaborative encoding your brain relates the new data with pre-existing knowledge to remember the facts in an efficient way is called Spaced Repetition. Just as the name sounds, spaced retrieval relies on your brain recalling information or knowledge at different intervals to help it quickly store and retrieve the memory without confusion. This technique is more commonly used with people who have neurological diseases that hinder their memory (such as Alzheimer's or Parkinson's). The method is rather simple: identify a target piece of information you wish to reliably recall. Have someone ask you the question, if you get it right the first time, then wait 15 seconds before asking the question again. If they get it wrong, repeat the process until they get it right, when they do get it right, double the time interval until you can give the right answer every time.

Perhaps the most basic yet effective type of memorization tool we humans know, mnemonics is where we assign anything that symbolizes or reminds us of the information we're meant to recall. For example, if I wanted to remember the first 100 digits of Pi, I could assign each digit a letter and make words using those letters that help me recall which digits come next.

Following on from Mnemonics, a mind palace is a complex yet most effective way to remember large quantities of information. The way it works is simple. First pick a place that you know pretty well, like your house or your school (that'll be your "palace"). Then based on what you want to remember you assign objects to them. Then as you "retrace" your steps through your palace you see those objects placed around in order of which you wanted to remember first. It is while hard to master, very rewarding if you manage to work it out.

Simonides is widely regarded as the historical inventor of the loci method. This is where you memorize large amounts of data can be remembered by placing images that represent the data into mental locations of a journey (essentially the mind palace technique). The story goes that Simonides of Ceos was dining at a nobleman's house with a lot of other people when he was called outside by two young men. While he was outside (this gets dark quickly), the roof of the hall where the banquet was being held collapsed and killed everyone underneath it. Burial parties were unable to identify the people due to their crushed and mutilated features, but Simonides was able to do so because he remembered where each person was sitting at the time of death. He would go on to suggest this practice of orderliness in memory recollection as an effective method.

Also known as smart drugs, or enhancers, nootropics are supplements that improve cognitive function. This includes motivation, memory, or creativity. Only a few drugs are known to improve some aspect of cognition, with most of them being stimulants, such as caffeine. The main two types of nootropics are racetams (such as piracetam, oxiracetam, and aniracetam) and central nervous system stimulants (such as caffeine, nicotine, amphetamine, and eugeroics). Most of them are available over-the-counter at pharmaceutical stores, yet the actual extent of their effects on the brain and negative side effects are still being explored.

Transcranial magnetic stimulation is a fairly simple concept though this may sound very complex and scientific. In transcranial magnetic stimulation, a patient is subjected to a magnetic field to assess the damage from stroke, multiple sclerosis, and other neurological and nervous-system-related injuries. During the process, a magnetic field generator is placed near the head of the patient, the coil is connected to a pulse generator that delivers a changing electric current to the coil. Recently, however, several studies have shown that TMS might have applications in increasing the long-term memory of patients by boosting the brain using the magnetic field. A total of 16 volunteers took part in an experiment by scientists at the Northwestern University of Chicago. Some of them were given TMS for 20 minutes a day, 5 days straight; others were told they were being given TMS but weren't actually (placebo condition). At the start of the test, they were shown 20 photos of human faces while hearing words associated with those faces being read aloud. After their treatment, they were asked to remember which words went with which face. The results found that those who'd received TMS had higher scores than those who didn't.

VIII

Terms to Know

Chunking is one of the most common techniques we use to remember information. In this process, we usually take data and process them into smaller parts called chunks to make it easier for our working memory to grasp. By clumping them into smaller parts they are classified as easy-to-remember bits. Thus making it easier for us to remember them over extended periods of time. Memorizing a phone number is the perfect example of this method at play. When we memorize a phone number we neither remember the number as a whole (i.e 9176321644) nor do we memorize each individual digit (i.e 9-1-7-6-3-2-1-6-4-4). Though you may not be doing this consciously, you actually clump the digits into groups(i.e 91-7632-1644)

Priming is a memory effect that influences our brain to associate a piece of information with another. Like every side of a coin has two sides, this effect too can have a positive and negative impact on a person. Positive priming occurs when you use the data to correlate two topics and this helps you retain the information. For example, when you think of the color yellow, you automatically think of a mango or a sun. This occurs because in your brain the words are 'primed' or are closely related and hence it's almost second nature for you to think of the other words, topics, phrases, or data linked to it. While positive priming enhances your ability to recall, negative priming generally slows it down. The mind can be negatively primed by exposing the person to various stimuli before ignoring these stimuli completely. It is believed that the brain deliberately sends a message saying to forget about ignored stimuli. When the brain tries to retrieve the ignored information, a conflict occurs. This conflict takes time to resolve resulting in negative priming

Yet another unexplained effect an event or trait has on human memory is the generation effect. The effect basically states that information is better recalled if it comes from your own brain as opposed to reading it. There are several explanations as to why this occurs, though not one of them has been found to be the definite answer.

Within the fields of psychology and cognitive science, the positivity effect is the ability to analyze a situation where the desired results are not achieved, yet positive feedback is obtained in order to help us become better in the future. It also states that our memory helps us develop more when we are given positive feedback instead of negative feedback.

Simply put, the humor effect states how memorizing something when some sort of humor is involved allows us to better retain it within our minds. A couple of experiments have been carried out about this, including one where a normal and dysphoric (dissatisfied) group watched humorous videos and were then able to recall more words from the clips. Scientists are still unsure as to what completely justifies this phenomenon, but some believe that the dopamine generated as a result of the happiness from the humor also stimulates long-term memory.

The Multi-Store Model says that the human memory is made up of 3 components. The first is the sensory register, where sensory information enters the memory. Short-term memory follows, where the inputs from the sensory register and retrievals from long-term memory are held for recollection. The final stage is long-term memory, where information rehearsed in the short-term memory is held for indefinite periods of time.

The level of processing effect is the direct opposite of the multi-store model. It was proposed that the depth of mental processing affected the way in which the memory functioned. Thus memories that are deeply processed are longer-lasting memories while those that do not receive the same level of processing are easily lost. The main two categories here are Shallow and Deep processing. The first of these occurs in four ways: Structural (processing how an object looks), Phonemic (How something sounds), Graphemic (Letters within a word), and Orthographic (the shape of something). Deep processing occurs in three ways: How an object or situation relates to something else, when we grasp the meaning of something and when the importance is comprehended. Beyond this, there are three main factors that determine if memory is retained for long periods: Maintenance Rehearsal (the process of repeating information), Elaborative Rehearsal (When the information is inspected to a greater extent), and Distinctiveness (when we are able to tell things apart).

Interference theory is the belief that memory can be disrupted or interfered with when what we have previously learned and what we're learning conflict with one another.

The idea suggests that information stored in long-term memory can become disrupted or confused when it combines with incoming information during the encoding process. It also states that memories are forgotten because of these interferences, thus leading to us forgetting things we thought we'd solidified in long-term memory. There are two main ways in which this can happen: proactive interference and retroactive interference. The first of these happens when you cannot learn a new task because of an old task you've already learned and your procedural memory has solidified. What we already know interferes with what we're already learning, thus causing the disruption and complete loss of memory. Retroactive interference occurs when you forget a previously learned task because of the one you're learning currently. The key thing to remember is that proactive interference is where old memories destroy new ones, while retroactive memory is when new ones disrupt old ones.

Within the field of psychology, memory inhibition is the ability not to remember irrelevant information. According to many scientists, this feature is a critical component of a memory system. The ability to lose memories when they're no longer useful is an adaptive trait, as it further hones in on efficient and rapid recollection. For example, if you're trying to remember your flight number for a trip you'd only want to remember that flight number and not every single flight you've been on in your life. In order to remember something, it is compulsory that we inhibit irrelevant information while we activate the relevant ones.

Essentially the Working Memory Model is the proposed theory in which short-term memory categorizes different types of information. The first of these is known as the Central Executive, the boss of working memory and sorts out which memories go to which subsystem. It also deals with cognitive tasks such as mental maths and problem-solving. Your visuospatial sketchpad, otherwise known as the inner eye, deals with storing or retrieving information in visual form (i.e helping you remember where you are in relation to objects in your environment, navigation). The final module of the working memory model is known as the Phonological loop, this deals with spoken and written material and you've probably used it when trying to remember a phone number. The Phonological store within the loop is where your memory holds information in a speech-based form (such as spoken words) for 1-2 seconds. The Articulatory control process controls speech production and is used to store and rehearse verbal information from the phonological store.

IX

When Memory goes Awry

Would it ever be ethical to change or remove someone else's memories? That depends on the situation. If the memory contains things you wouldn't want them to be burdened with (i.e memories of a painful time or the loss of something dear to them). If however the loss or manipulation of said memory is only for our benefit, then it doesn't make sense to erase it.

Is there any way to know for sure whether our memories are accurate? As of now, no. Eventually, science and technology will develop a machine or module that can be inserted into our brain to take photos of memories we wish to store. Our mind can create an illusion and have a bias thus altering the actual memory

How accurate is eyewitness testimony? Can it be misleading, and can it be improved? Eyewitness testimony is one of those things that's like a paradoxical situation from which an individual cannot escape because of contradictory limitations. How do you get better witnesses from a crime scene that has only those eyewitnesses? Eyewitnesses can be very misleading, the amount of time between the event and the questioning can often allow the memory to be manipulated or harder to retrieve. what the eye sees is perceived differently from person to person. That's why detectives immediately try to place this job first so that the memory is still "fresh" in their brain. It can certainly be improved, by training people from birth to spot the smallest details and include them in their memories.

Do people from different cultures and societies remember the same things differently in predictable ways? This is proven in several studies. People in individualistic cultures are more likely to remember the same things more vividly and accurately as opposed to those in collectivistic cultures.

To what extent can we trust decisions made by those without sound memory? Not a lot honestly, as without sound memory they could repeat their mistakes again and not learn from them though it is not their fault. Especially if their decisions are crucial in deciding the result of something, say an investigation or a key choice that might change everything.

X

On the Tip of the Tongue

Decay Theory states that the simple passage of time causes our memories to fade. The more time that passes between the creation of the memory and when we need to recall it, the harder it will be for us to retrieve it. Neurons are activated when any information enters the brain and those memories stay in our heads as long as the neurons are active. Frequent recall of the information and rehearsal are two ways to keep the neurons active. But, if the activation isn't maintained, the memory decays and fades.

Gaslighting refers to a form of manipulation in which an individual or a group of individuals sow seeds of doubt targeting others with a different perception of things in such a way that they end up questioning their own memory or judgment.

Motivated forgetting is a theorized psychological behavior in which people may forget unwanted memories, either consciously or unconsciously. Although it might get confusing for some, it's completely different from defense mechanisms. Motivated forgetting is also defined as a form of conscious coping strategy. For instance, a person might direct his/her mind towards unrelated topics when something reminds them of unpleasant events. This could lead to forgetting of memory without having any intention to forget, making the action of forgetting motivated. It is of two types- Conscious and unconscious.

Psychological Repression, an unconscious act

The concept was based on Sigmund Freud's psychoanalytic model, which suggested that people subconsciously push unpleasant thoughts and feelings into the unconscious. However, repressed memories, although repressed, have been known to influence behavior, dreams, decision making, emotional response, and so on. For instance, a child abused by a parent, who had repressed the memory, has trouble forming relationships. Psychoanalysis was the treatment method offered by Freud for repressed memories, with the goal to bring back the fears and emotions to the conscious level.

Thought Suppression, a conscious act

The deliberate or conscious attempt to suppress memories is referred to as thought suppression. This phenomenon involves conscious strategies and intentional context shifts, so it is goal-directed. For instance, if a person faces stimulants of unpleasant memories, he/she might deliberately try to push the memory into the unconscious by thinking about something else. But, thought suppression can be a time-consuming task and quite difficult too. Also, the memories can easily resurface with minimal prompting, which is why it's closely associated with Obsessive-Compulsive Disorder.

TBI or Traumatic Brain Injury occurs when an external force injures the brain. This injury can be permanent or recoverable, widespread or closed, and penetrating. The consequences can be severe, from permanent disability to death. Within the field of memory, however, TBI can cause damage to your long-term or short-term memory. If your amygdala or hippocampus is damaged severely enough, then it is possible amnesia will form and you'll be unable to remember much about your life before the traumatic injury was delivered.

An amnesia is a dramatic form of memory loss. But it is far more complicated and severe than everyday forgetfulness. Forgetting what your spouse asked you to pick up at the grocery store is "normal." Forgetting that you are married can be a sign of amnesia.

Amnesia is often portrayed in movies and TV shows. What soap opera hasn't had a storyline involving it? Fictional characters with amnesia often lose their entire identities. They can't even remember their names. Fortunately, amnesia usually isn't that severe in real life.

If you have amnesia you may be unable to recall past information (retrograde amnesia) and/or hold onto new information (anterograde amnesia). Both can occur simultaneously.

Post-traumatic Amnesia is the amnesia that occurs immediately after a significant head injury. It may involve retrograde amnesia, anterograde amnesia, or both.

Transient Global Amnesia is A temporary syndrome where you experience both retrograde and anterograde amnesia. Memory loss is sudden and only lasts up to 24 hours.

Infantile Amnesia is the term used to describe the fact that people can't recallmemories of events from early childhood. Few people have memories from before the ages of three to five because the brain areas that support memory are still developing.

Dissociative Amnesia/Psychogenic Amnesia is A mental health disorder where you experience amnesia after significant trauma. You block out both personal information and the traumatic incident from your memory

Unlike the past conditions listed here, blackouts are only temporary and are instead caused by something far less mortifying: alcohol (insert sarcastic gasp here). Blackouts occur when your body's blood alcohol content is high, usually above 14 percent (it can differ depending on age, weight, and other biological factors). While your blood alcohol content is above this amount, you may not have any memory about the time that's passed since it peaked that safety cap of 14 percent.

Ever had that feeling of wishing someone long gone was by your side, or that you were still back in that peaceful home? That's when you've experienced nostalgia. Nostalgia is essentially when your brain yearns to return to a former period or for someone to return to your life. Often triggered when lonely or by memory cues, it can bring a sense of close relationship and communal bonding.

You've probably already heard of this one. A flashback occurs when an individual has a powerful re-experiencing of a past moment in their lives or elements of it. Unlike normal memory, flashbacks happen involuntarily and are so intense that the person is said to "relive" the experience.

Post Traumatic Stress Disorder, commonly known as PTSD is triggered when one experiences an event that traumatizes them. They needn't necessarily be the ones experiencing it but even witnessing such a situation can induce PTSD. Symptoms usually include flashbacks, nightmares, and even severe anxiety. The people who receive PTSDs have usually gone through a lot in their lives, including refugees, soldiers, and survivors of disasters. If PTSD goes untreated, it can often result in a severe drop in the positivity of the person, they might not be the person you once knew.

Dissociation or Dissociative Identity is probably one of the more severe and harmful conditions on the list. Dissociation is essentially when you retreat from society, the real world, and emotion itself. It is a disconnection from your thoughts, feelings, actions, and sense of who you are.

There are three main types of dissociative disorders: Dissociative Identity, Derealization/Depersonalization, and Dissociative Amnesia. The first is what it says, you are not yourself, two or more different identities exist within you. The second is where you detach yourself from your own body and reality, you feel as if you are watching events happening around you. The final one is where you have trouble remembering things, from moments that recently occurred to even entire years of your life.

Now we're getting into the more disease-oriented areas of memory. Alzheimer's is probably the most well-known neurological condition there is and with good reason. It is a chronic disease, meaning that as time passes, the consequences worsen. The most common symptoms in Alzheimer's are related to lack of memory storage or the inability to host memories. The earliest symptoms include difficulties in remembering recent events (short-term memory). As the disease begins to develop, the severity of the conditions means that a patient with Alzheimer's is usually only going to live for at most 10 more years. The more severe conditions include language problems, disorientation, mood swings, loss of motivation, not managing self-care, and other behavioral issues. In the final stages of the development, the disease begins to cause your mind to lose knowledge of body functions, eventually leading to death. The reason why Alzheimer's is so infamous is because of how little we know about it. Doctors and psychologists aren't even sure what causes it (the best guess is genetic mutation) and no treatment exists to stop or reverse the condition.

Directly linked to Alzheimer's, dementia is not so much a specific condition but instead refers to a wide range of symptoms linked with a decline in memory or complete removal of thinking skills that might limit a person's everyday functions. Alzheimer's is the most common cause of dementia, accounting for 60 to 80 percent of causes.

Now here is a condition we have far more information about than Alzheimer's. Korsakoff's syndrome is a chronic memory disorder caused by a lack of thiamine otherwise known as Vitamin B-1. This essential nutrient helps brain cells to produce energy from sugars and a lack of it can cause brain cells to stop functioning, leading to problems with memory on all levels. Those problems include difficulties acquiring new skills, remembering recent events, and long-term memory gaps. While you might be able to carry on a conversation with a person, you might forget just minutes after the conversation ends what it was about or who you talked to. Doctors recommend that those who are heavy drinkers (for alcohol misuse is a common cause of Korsakoff's) take an oral supplement of thiamine and other vitamins.

XI

Biases and Fallacies

The seemingly natural disruption in a cognitive process when you sense the need to preserve and increase self-esteem or praise oneself is known as self-serving bias. When you give yourself full credit for your accomplishments while blaming failure on any external circumstances.

The phenomena of rosy retrospection occur when people look back on their past with disproportionately greater fondness than they do in their present. It relates to steroid-induced nostalgia and hallucinogenic drugs (literally, it is only different in the fact that it is a cognitive bias). The Romans used the word "memoria praeteritorum bonorum" to describe this phenomenon (The past is always well remembered).

Confabulation is a memory disorder in which you recall or create false, distorted, or misunderstood recollections about yourself or the environment. The fact that you did not intend this memory to be deceptive and so did not intentionally modify it sets it different from other words for false memories (and keeps it from being a sin).

It's time to pick up another language. Psychological repression, also known as Verdrängung in German, is a purposefully performed psychological process in which you push unpleasant ideas, emotions, and desires into the subconscious world, where they cannot be accessible. for example, When you receive very bad news, you may wish to file it away for later reference or forget about it entirely.

Repressed memories do not vanish altogether; rather, they fade from our consciousness until they are reactivated. Repressed memories might manifest themselves in your actions, such as worry or dysfunctional conduct. They can also be let out in nightmares or through a slip of the tongue also known as Freudian slips.

Memory implantation, which is more of a method than a condition, is used to study human memory and includes psychologists convincing you of something that never happened. The more believable false memories that have been implanted are generally the ones that have been implanted effectively.

Memory conformity is the psychological equivalent of jumping on the bandwagon. Memory conformity occurs when your mind changes a recall of an event after hearing another group/recollection person of the same event or experience. Due to societal pressures, it is deemed a mistake and is highly troublesome for legal procedures, daily memory, and investigative work. Have you ever wondered why just one eyewitness is called in to be questioned? You now understand why. Even a small change can mean a badly done investigation or a criminal not being brought to justice.

Your mind can be temporally shifted when an event occurs, similar to how a telescope makes objects appear closer or farther away than they are. Backward telescoping occurs when your brain believes a recent occurrence happened a long time ago. When your brain does the opposite and thinks that a distant event occurred more recently, that's known as forwarding telescoping.

This umbrella encompasses several of the previous terms. Recall bias is a type of memory mistake in which there is a discrepancy in recollection accuracy or completeness. Although memory bias is most commonly used in medical areas to allude to a patient's inability to provide sufficient or accurate information to make a diagnosis, it is also quite frequent in our everyday lives.

The seven memory sins are those that you should avoid committing if you had superior memory and storage.

The first is transience, which occurs when a person's memory deteriorates over time. While this is generally typical as you get older, severe versions of it can be caused by degradation or damage to the brain's memory modules (i.e. the hippocampus).

The second sin is absent-mindedness, which occurs when attention and memory collide. Examples of this type of mistake include forgetting where items are stored or failing to make appointments on time.

The third sin is blocking, which occurs when your brain wants to store or retrieve information but is blocked by another memory. This is the primary cause of the Tip of the Tongue Phenomenon, in which you are briefly unable to retrieve stored knowledge.

The fourth sin is a misattribution, which occurs when you properly recall knowledge but don't know where it came from (for example, when you remember a specific piece of information but can't remember which documentary it came from). This sin is particularly troublesome in court proceedings and professions that rely on eyewitnesses to do their jobs better.

The fifth sin is suggestibility, which occurs when you accept a false suggestion that causes your memory and recollection to be altered.

The second final sin is simply called bias, and it's similar to suggestibility in that it involves focusing on one element of your life or a memory rather than the full picture. Looking back on all of the pleasant feelings you experienced as a child without addressing the bad ones, for example, may lead you to conclude that your childhood was a happy one.

Persistence is the last sin, a memory failure that happens when your brain repeatedly recalls undesirable memories or knowledge, which can result in significant emotional effects such as phobias, PTSD, and even suicide.

XII

Technologies of Remembrance

Can technology help us to remember things? They already are. Treatments like nootropics and transcranial magnetic stimulation are already technologies that enhance our brain's ability to retain and recall information. As of now, there isn't one definitive, proven technology that can allow us to store every detail of memory and recall it perfectly after a long period. Yet I suppose it's not illogical to say that shortly (if we even last that long), such technology will exist.

Can they help us to forget them? While it is the work of many science-fiction authors and movie directors alike to explore brainwashing and memory erasable machines, a more basic one exists. Technology is, by being a database of virtually any information, easing the strain on our memory to hold that information. In a previous generation, if you saw a fact in a book, you'd probably need to cram your memory with that fact or risk having to search for the book and page all over again. Now, instead of having to memorize a fact, we simply pull up the nearest piece of technology and search it up on the internet.

How ironic that mankind's march forward into the future may also mean a step backward for historical signs of progress. Simply put, the digital dark age is the fear shared by many historians and archivists that we will lose records of the past because computers in the future will be unable to read the data stored in our current ones. There is such a fear that many claim we may one day "know more about the 20th century than the early 21st century" because the lack of physical records allows many problems of recollection to arise

Obsolescence is a term used to describe any product, service, or software that is no longer used even though it may be in perfect working order. Even though your new iPhone 8 might be working perfectly and is pretty much brand new, Apple will make sure that the iPhone X makes your smartphone obsolete (mostly due to purposeful software degrades). This is becoming a huge problem in businesses, where they purposefully plan obsolescence so that you have to go buy new products, thus generating more profit for them.

Standing for Random-access memory, RAM is a form of computer data storage that stores instructions and data currently being used. In comparison to other storage devices like hard-disk drives and drum memory, RAM allows data to be read or written at almost the same time regardless of where the data is within the memory.

Eventually, a piece of software has outlived its time or can no longer be used, instead of deleting the code completely, some manufacturers and owners simply leave it to decay in the dark abyss of computer software. No longer accessible due to a lack of support, the copyright infringements of such software generally no longer exist.

This is termed Abandonware.

Digitization converts information into a computer-readable format by converting it into binary code. Almost all media can be converted this way.

Drum memory was the precursor to the hard disk drive that now occupies most computers or laptops. It was a magnetic data storage device shaped like a drum barrel and was primarily used as the main working memory of computers. The prototype of drum memory was only able to hold about 62.5 kilobytes, whereas modern hard disk drives in pre-built desktops can hold 523,763,720 kilobytes.

Punch cards are paper cards where holes may be punched by hand or machine to represent computer data and instructions. They were a widely-used means of inputting data into early computers. The cards were fed into a card reader connected to a computer, which converted the sequence of holes to digital information.[14] For example, an early computer programmer would write a program by hand, then convert the program to a series of punched cards using a punch card machine. The programmer would then take the stack of cards to a computer and feed the cards into a card reader to input the program. Pictured is an example of a woman using a punch card machine to create a punch card.

One of the very first examples of smart automation within a workplace, the Jacquard loom was a machine devised by Frenchmen Joseph-Marie Jacquard to allow looms to produce complex patterns on fabric. It was essentially a paired device, designed to use punch cards (cards with holes punched in specific places that meant different things) that controlled the weaving of the cloth so the desired pattern could be created. See the picture below for an illustration. The mechanism of this machine would lay the groundwork (or should I say programming?) of English inventor Charles Babbage's analytical engine and early computers.

A Stored-program computer was a concept that stated that a computer memory with stored instructions inside it could perform a variety of tasks in sequence or intermittently. Neumann proposed that a program could be electronically stored in a memory device using the binary-number format. It allowed digital computers to become much more flexible and powerful and was successfully implemented in England through the Manchester Mark I, the first stored-program computer.

A crawler, also known as a web-crawler or web spider, a crawler is an internet bot that systematically browses programs that visit Web sites and reads their pages and other information to create entries for a search engine index. The major search engines on the Web all have such a program, which is also known as a "spider" or a "bot." Crawlers are typically programmed to visit sites that have been submitted by their owners as new or updated. Entire sites or specific pages can be selectively visited and indexed. Crawlers gained the name because they crawl through a site a page at a time, following the links to other pages on the site until all pages have been read.

Metadata is a whole other layer of data regarding the data itself. It essentially describes a single piece of data or a collection of it for ease of searching and retrieval. For a text, this metadata might include the author, publishing date, title, and the number of pages. Generally, metadata can be automatically generated by programs or manually created using meta tags, descriptions, and keywords to describe their content and help search engines locate them.

Gamers and printer enthusiasts are probably familiar with this. Emulation is the process that an emulator undertakes. An emulator is a piece of software or a program that allows one computer system known as the host to act like another known as the guest. Emulators allow the host system to run software or use peripheral devices designed for the guest system. Emulation refers to the ability of a computer program to emulate or imitate another device. There's a reason why many printers are designed to emulate Hewlett-Packard LaserJet ones because most software is written for them. Any software written for an HP printer can then be used by the emulated printer to produce equivalent results in printing. Video games also have this, with many enthusiasts using emulators to play games on different platforms than they were intended to. Other emulators allow one to play much older games from the 80s or 90s that can't be found anymore online.

Caching is a technique that stores a copy of a given resource and serves it back when requested. When a web cache has a requested resource in its store, it intercepts the request and returns its copy instead of re-downloading from the originating server. This achieves several goals: it eases the load of the server that doesn't need to serve all clients itself, and it improves performance by being closer to the client, i.e., it takes less time to transmit the resource back. For a website, it is a major component in achieving high performance. On the other side, it has to be configured properly as not all resources stay identical forever: it is important to cache a resource only until it changes, not longer.

There are several kinds of caches: these can be grouped into two main categories: private or shared caches. A shared cache is a cache that stores responses for reuse by more than one user. A private cache is dedicated to a single user. This page will mostly talk about browser and proxy caches, but there are also gateway caches, CDN, reverse proxy caches, and load balancers that are deployed on web servers for better reliability, performance, and scaling of websites and web applications.

XIII

Disorders

Memory comes in a variety of shapes and sizes. We all know that when we store a memory, we're actually storing data. However, the sort of memory we have is determined by the material we remember and how long we remember it. Short-term memory and long-term memory are the two most common types of memory, based on how long the information is maintained. Both can deteriorate as a result of aging, as well as a number of other factors and medical disorders that influence memory.

Schizophrenia is a severe mental illness in which patients have distorted perceptions of reality. It can include hallucinations, delusions, and severely disorganized thought and behavior, which can make it difficult to operate on a daily basis. Schizophrenia patients need to be treated for the rest of their lives. Early therapy can assist to manage symptoms and improve the long-term outlook by preventing severe consequences. Symptoms of schizophrenia include delusions, hallucinations, disordered speech, difficulty thinking, and a lack of desire. Despite the fact that there is no cure for schizophrenia, research is leading to new and safer therapies. The disease's intricacy may explain why there are so many misunderstandings about it. Schizophrenia does not imply multiple personalities or divided personalities. The majority of persons with schizophrenia are no more dangerous or violent than the general public. While a lack of community mental health supports may result in homelessness and repeated hospitalizations, it is a common misunderstanding that persons with schizophrenia end up homeless or in hospitals. The majority of persons with schizophrenia live with their families, in group homes, or alone.

Autism, often known as autism spectrum disorder, is a term used to describe a group of disorders marked by difficulties with social skills, repetitive activities, speech, and nonverbal communication. Autism spectrum disorder encompasses disorders such as autism, Asperger's syndrome, childhood disintegrative disorder, and an undefined kind of pervasive developmental disease that was previously thought to be distinct. Some individuals still refer to autism spectrum disorder as "Asperger's syndrome," which is considered to be on the milder end of the spectrum. Autism spectrum condition manifests itself in early life and leads to difficulties in social, educational, and occupational settings. Autism signs appear in youngsters as early as the first year of life. A tiny percentage of children appear to grow normally in the first year, but subsequently regress between the ages of 18 and 24 months, when they begin to show signs of autism.

The unjustified and persistent feeling that others are out to get you or that you are the object of others' continual, intrusive attention is known as paranoia. Paranoiacs may find it challenging to function socially or sustain close relationships because of their unreasonable dread of others. Paranoia can be a sign of a variety of disorders, including paranoid personality disorder, delusional disorder, and schizophrenia, among others. The etiology of paranoia is unknown, however, it is considered that heredity may play a part. Treatment is determined on the underlying problem and may involve psychological counseling or medication.

Anxiety is a state of mind marked by tense sensations, anxious thoughts, and physical changes such as elevated blood pressure. Anxiety disorders are characterized by recurrent intrusive thoughts or concerns. They may avoid certain situations because they are concerned. Physical symptoms such as sweating, trembling, disorientation, or a fast heartbeat may also be present.

Insomniacs have trouble falling asleep, staying asleep, or getting enough restorative sleep. Insomnia is a sleep condition that affects many people. Sleep deprivation can lead to health issues such as diabetes, hypertension, and weight gain over time. Behavioral and lifestyle adjustments might help you get a better night's sleep. Sleeping medications and cognitive-behavioral therapy can also assist. Women are more likely than males to suffer from insomnia. Sleep can be disrupted by pregnancy and hormonal changes. Sleep can also be affected by hormonal changes such as premenstrual syndrome or menopause.

A panic attack is an overwhelming surge of anxiety that comes on suddenly and is devastating and immobilizing. You can't breathe, your heart is pounding, and you may feel like you're dying or going insane. Panic episodes often happen out of nowhere, without warning, and with no obvious cause. They can happen even while you're relaxed or sleeping. A panic attack might be a one-time occurrence, but many people have them more than once. A specific circumstance, such as crossing a bridge or speaking in public, can often provoke recurrent panic attacks, especially if the situation has previously sparked a panic attack. The panic-inducing circumstance is usually one in which you feel threatened and helpless, prompting the fight-or-flight reaction in your body.

A phobia is a persistent, irrational dread of something, someone, an animal, an activity, or a circumstance. It's an anxiety condition of some sort. A person who has a phobia either attempts to avoid or suffers the item that causes the dread with significant worry and anguish. The term phobia is frequently used to describe a dread of a certain trigger. There are three different kinds of phobias:

- Specific phobia is an unreasonable dread of a certain trigger.
- Social phobia is a strong dread of public humiliation and being singled out or evaluated by others. For someone with social anxiety, huge social gatherings are scary. It is not to be confused with shyness.
- Agoraphobia is a dread of circumstances from which it would be difficult to escape if a person were really panicked, such as being in an elevator or being outside of one's house. It's sometimes misconstrued as fear of wide spaces, although it may also refer to being trapped in a tiny area, such as an elevator, or on public transportation.

Bipolar disorder, often known as manic depression, is a mental illness characterized by severe mood swings and emotional highs and lows. You may feel gloomy or hopeless when you are depressed, and you may lose interest or pleasure in most activities. You may feel ecstatic, full of energy, or particularly irritated when your mood swings to mania or hypomania. Sleep, energy, activity, judgment, conduct, and the capacity to think clearly can all be affected by mood fluctuations. Mood swings might happen once a year or several times a year. While the majority of people will feel some emotional symptoms in between bouts, others will not. Although bipolar illness is a lifelong diagnosis, following a treatment plan might help you control your mood swings and other symptoms. Medications and psychological counseling are used to treat bipolar illness in the majority of instances.

An illness that affects millions of children and often lasts into adulthood, attention-deficit/hyperactivity disorder is a chronic condition. ADHD is characterized by a number of persistent issues, including the inability to maintain focus, hyperactivity, and impulsive conduct. Children with ADHD may also have low self-esteem, strained relationships, and poor academic achievement. Symptoms may diminish as you become older. Some people, however, never fully recover from their ADHD symptoms. They can, however, develop successful techniques. While therapy will not cure ADHD, it can significantly reduce symptoms. Medications and behavioral treatments are commonly used in treatment. Early detection and treatment can have a significant impact on the result.

Seasonal affective disorder (SAD) is a kind of depression that is linked to seasonal variations. It starts and finishes around the same time every year. If you're like the majority of SAD sufferers, your symptoms begin in the fall and last through the winter, sapping your energy and making you irritable. In the spring and early summer, SAD is less likely to induce depression. Light treatment, medicines, and psychotherapy may be used to treat SAD.

Depression is a type of mood illness characterized by a continuous sense of melancholy and a loss of interest. It affects how you feel, think, and behave and can lead to a range of mental and physical issues. It's also known as major depressive disorder or clinical depression. Many persons with depression have significant symptoms that interfere with their day-to-day activities, such as job, school, social activities, or interpersonal relationships. Some people may be sad or unpleasant in general without knowing why.

Dyslexia is a learning disability in which people have trouble reading because they can't recognize spoken sounds or understand how they connect to letters and words. Dyslexia, often known as reading impairment, affects the parts of the brain that process language.

When the blood flow to a portion of your brain is stopped or decreased, brain tissue is deprived of oxygen and nutrients, resulting in a stroke. Within minutes, brain cells begin to die. A transient ischemic attack— sometimes called a ministroke — is a brief period of symptoms that resemble those of a stroke. A transient ischemic attack has no long-term consequences. They're triggered by a brief interruption in blood flow to a portion of your brain, which can last as little as five minutes. A TIA happens when a clot or debris decreases or stops blood flow to a portion of your neurological system, similar to an ischemic stroke.

Munchausen syndrome is a mental illness in which you fabricate, exaggerate, or intentionally cause physical, emotional, or cognitive problems. People with factitious diseases behave in this way to satisfy an inner desire to be perceived as sick or damaged, rather than to receive a tangible advantage, such as medicine or financial gain. This is distinct from malingering, which occurs when someone exaggerates or fabricates an ailment in order to avoid work.

Psychosis is a mental illness that alters how information is processed in the brain. You lose contact with reality as a result of it. It's possible that you'll see, hear, or believe things that aren't true. Psychosis is a symptom, not a medical condition. A mental condition, a physical injury or sickness, substance addiction, or acute stress or trauma can all induce it.

ɞ

Locked-in syndrome is a rare neurological disease in which all voluntary muscles are paralyzed except those that govern the eyes. People with locked-in syndrome are aware of their surroundings and can reason and think, but they are unable to talk or move. Blinking and vertical eye motions can be utilized to communicate.

Brain stem stroke, traumatic brain damage, tumors, circulatory system illnesses, diseases that disrupt the myelin sheath covering nerve cells, infection, or drug overdose can all produce locked-in syndrome. The locked-in condition has no cure or particular therapy. Supportive breathing and feeding treatment are critical, especially early on. The foundation of treatment is physical therapy, comfort care, nutritional assistance, and the avoidance of systemic problems including respiratory infections.

While some muscular control recovery is feasible, it is partly reliant on the underlying reason. With eye movements and blinking, speech therapists can help patients with locked-in syndrome communicate more clearly. Electronic communication technologies, such as infrared eye movement sensors and computer voice prostheses, are helping persons with locked-in syndrome to interact more freely and use the internet. Due to medical problems, many persons with locked-in syndrome do not survive through the early stages. Others, despite the severe impairments produced by the condition, may survive for another 10-20 years and report a decent quality of life.

XIV

Memory Development

Restoring Active Memory

The Restoring Active Memory initiative seeks to help military service members adapt to the impacts of traumatic brain injury by discovering neurotechnologies that enhance memory formation and recall in the damaged brain. Since 2000, more than 270,000 service members have been diagnosed with TBI.

The condition usually causes an impairment in the ability to recover memories created previous to the damage, as well as a diminished capability to develop or store new memories after the injury. Despite the severity of the problem, there are presently few viable treatments for reducing the long-term effects of TBI on memory. Enabling memory function restoration would boost military readiness by allowing injured troops to return to duty, as well as improving the quality of life for wounded veterans.

Performer teams are developing multi-scale computational models with a high spatial and temporal resolution to explain how neurons code declarative memories, which are well-defined chunks of knowledge that can be consciously recalled and articulated in languages, such as events, times, and locations. Teams are also looking at novel ways for analyzing and decoding neural signals in order to figure out how focused stimulation may be used to assist damaged brains to regain function.

Building on this foundational work, researchers are integrating the computational models into new, implantable, closed-loop systems able to deliver targeted neural stimulation to restore normal memory function. RAM also funds animal research to enhance the state-of-the-art of quantitative models that account for the storage and retrieval of complex memories and memory characteristics, as well as their hierarchical relationships.

HDAC inhibition

Scientists are working on novel techniques to enhance gene expression in the brain preferentially in order to diagnose psychiatric and neurological disorders. Medications that target this pathway have been shown to aid learning and memory in rats, as per a growing set of knowledge. Existing medicines, however, which were not created for this purpose, are weak and beta-agonist, and their long-term safety is uncertain.

In current history, neuroscientists have begun to understand the relevance of epigenetics in the brain, particularly in memory. Epigenetics are chemical mechanisms that modify the expression of genes without altering DNA.

Histone deacetylases (HDACs), a group of enzymes that cause DNA to coil more tightly around adjacent proteins, suppressing gene expression, are one of the most important epigenetic regulators.

Existing medicines that block these enzymes have been demonstrated in recent research to improve learning in both normal and cognitively impaired animals.

Last year, a researcher at MIT discovered that giving brain-damaged rats an HDAC inhibitor helped them to remember memories they had previously forgotten. EnVivo Pharmaceuticals, located in Watertown, Massachusetts, is working on HDAC inhibitors that are more powerful than current ones and can easily reach the brain.

The company's main HDAC inhibitor can improve both short- and long-term memory in mice, according to data presented at a neuroscience conference last month. While scientists aren't sure how epigenetic regulation impacts memory, one idea is that certain stimuli, such as exercise, visual stimulation, or medications, unwind DNA, allowing genes involved in brain plasticity to be expressed.

This increase in gene expression might lead to the creation of new synaptic connections, which would enhance the memory-forming neural circuits.

Perhaps these epigenetic pathways are used by our brains to allow us to learn and recall things, or to offer substantial versatility to allow us to learn and adapt.

HDAC inhibitors have been shown to induce extensive dendritic development and enhance synaptogenesis [the generation of connections between neurons].

By reorganizing or mending damaged brain circuits, the procedure may improve memory or allow mice to reclaim lost memories.

Optogenetics

There are very few things that science can do to reverse diseases like addiction, other than detox and a long, grueling recovery procedure.

Researchers have been able to rewrite pleasant memories linked with cocaine in drug-addicted mice. This discovery might provide light on how science can be utilized to neurologically change long-standing negative habits. The study's cognitive scientists proceeded by using cocaine to instruct mice to pick a certain area.

They utilized a genetic technology called optogenetics, in which living brain cells are manipulated via fiber optic cables, to change the mice's favorable connection with it. The mice effectively lost their preference for the cocaine-associated environment after the therapy, suggesting that the therapy allowed them to rewrite their memories.

During training, the mice were offered the choice of a saline solution or cocaine-laced milieu. The mice, predictably, would prefer the cocaine-related environment. After that, the researchers identified the cells in the mice's brains that were active in the cocaine-linked environment, causing the cells to produce light-sensitive proteins. Scientists 'switched on the light,' which silenced the tagged neurons, while the mice were meandering about the cocaine surroundings.

Unmasked neurons were activated as a result of the process, resulting in the formation of a new cognitive map. As a consequence, the mice lost their preference for the cocaine-associated location, indicating that memories are retained in the brain as a result of biophysical or biochemical changes induced by external stimuli. It also indicates that optogenetics might be used to rewrite harmful behaviors like addiction.

Creating false memories

A mouse was placed in a tiny metal box with a black plastic floor by Steve Ramirez, a 24-year-old doctoral student at the time. Instead of sniffing about with interest, the animal froze in fright, recalling its previous encounter with a foot shock in the same box. The mouse's stance was stiffer than Ramirez had expected, and it was a typical panic reflex.

Its trauma must have stayed with it for a long time. Which was incredible, because the recollection was false: the mouse had never been shocked in that cage.

Rather, it was reacting to a fake memory implanted in its brain by Ramirez and Liu. The discovery confirmed an astonishing hypothesis: not only could brain cells involved in the encoding of a single memory be identified but those cells could also be manipulated to generate a completely new "memory" of an event that never happened. The duo used a mouse to implant a fake memory in a neuroscience breakthrough.

For years, scientists have been enthralled by the idea of precisely manipulating memory. A lot of people have thought along similar lines but they never imagined that these trials would actually work. Their study has ushered in a new age in memory research and may one day lead to novel therapies for medical and mental disorders such as depression, post-traumatic stress disorder, and Alzheimer's disease. Despite the fact that the research has so far only been done on lab mice, the findings suggest a better understanding of human nature. What does it mean to have a past if memories can be changed at will? How can we establish a real sense of self if we can delete or create a vivid imagination?

ॐ

Is it possible to help someone who is depressed by reactivating good memories? And we mostly found out how to eliminate things like smallpox, which you can't see, and have to infer its existence from indirect data until your microscopes get good enough. The flashing clusters of neurons known as engrams, where individual memories are stored, have been seen and controlled by Ramirez and Liu.

They created a complex new approach for studying living brains in action, combining traditional molecular biology with the burgeoning science of optogenetics, in which lasers are used to activate cells genetically modified to be light-sensitive. They found, tagged, and then reactivated a tiny cluster of cells encoding a mouse's fear memory, in this case, a recollection of an environment where the animal had received a foot shock, in the first experiment.

The achievement backs with the long-held idea that memories are stored in engrams.

The majority of prior studies focused on the chemical or electrical activity of brain cells during memory formation. The scientists used a particular breed of genetically modified lab mice to inject a biochemical cocktail into the dentate gyrus that includes a gene for the light-sensitive protein channelrhodopsin-2. Active dentate gyrus cells, which are involved in memory formation, would generate the protein, making them light-sensitive.

The concept was that when the memory had been encoded, the cells could be zapped with a laser to reawaken it. To accomplish so, the researchers surgically inserted tiny laser filaments through the mice's heads and into the dentate gyrus. The only way to verify they had recognized and classified an engram was to reactivate the memory—and its accompanying terror reaction.

After the experiment, the animals were killed, and the brain tissues were examined under a microscope to establish the presence of the engrams; cells implicated in a specific memory shone green after being treated with substances that interacted with channelrhodopsin-2. Even though these activated cells were only a small fraction of a larger foot shock engram, reactivating them was enough to cause a fear reaction. The next stage was to construct a fake memory by manipulating a certain engram. They started by injecting the biochemical mixture into the dentate gyrus of the mouse. The mouse was then placed in a box without being shocked. A memory of this pleasant event was stored as an engram while the animal spent 12 minutes investigating.

The mouse was placed in a new box the next day, and the memory of the original (safe) box was triggered by firing the laser into the dentate gyrus. A foot shock was delivered to the mouse at that precise instant. The mouse was returned to the secure box on the third day and immediately froze in fright. It had never experienced a foot shock there, but it acted as though it had because of a fake memory generated by the researchers in another cage. There was no way the mouse could have mixed up one box with the other: They came in a variety of forms, colors, and fragrances.

The potential clinical uses of memory manipulation are tempting at a time when therapies for many major mental disorders are scarce. "However, someone with Alzheimer's disease." Perhaps we can devise a therapy that involves going in and doing what these researchers did in their publications, i.e., artificially stimulating these cells to enhance activation and improve memory recall.

In another possible application, PTSD might be alleviated by continually reactivating a terrible memory to demonstrate that the memory is not dangerous in and of itself, or by deleting the traumatic components of a specific negative memory, or by substituting it with a happy memory. Because we've proven that we can artificially reawaken memories and manufacture false memories in animals, the only thing standing between us and humans is technical advancement.

Neuroprosthetic implants

Motor, sensory, and cognitive functions that have been compromised as a result of nervous system diseases can be replaced by neuroprosthetic devices. Cochlear implants for individuals with hearing loss and prosthetic devices for amputees have been the most effective neuroprosthetic devices produced to date. Because both types of devices are on the outskirts of the nervous system's sensorimotor processing, replacing them was relatively easy, but the capabilities of certain consumer goods were restricted.

For paretic patients, neuroprosthetic devices have been created with the goal of not only restoring but also rehabilitating motor function. Researchers use a technique known as a brain-machine interface to capture signals directly from the brain and connect them to effectors.

Biosignals have been captured noninvasively or invasively to use such technology for neuroprosthetic devices.

The primary motor cortex has traditionally been the primary source of motor-related data, but a new study looks into the recording of signals from the middle frontal gyrus. To restore motor function, more sophisticated sensorimotor information from the brain might be retrieved. Because communication is strongly related to body part motions, patients who lose their voluntary movements also lose their methods of communication.

XV

Brain-Computer Interface

One of the most amazing things in the cosmos is the human brain. It is the body's most sophisticated organ and the source of our knowledge of the world. It's what permits you to read this text while simultaneously analyzing your environment and eating a snack. Processing in the brain is responsible for everything you touch, taste, feel, see, and hear. Most significantly, compared to other physiological systems, our understanding of the brain is minuscule: our circulatory, respiratory, endocrine, digestive, and other biological systems have all been mapped; our brain, on the other hand, has not. That's not even close.

What's more intriguing is that there's no knowing what a greater knowledge of the brain may lead to. One effect is the treatment of neurodegenerative disorders, although it is far from the only one. We may begin to rethink what it means to be human as our understanding of the brain evolves. BCIs (brain-computer interfaces) is a new technology that links the human brain to a computer or machine (BMIs) and interprets brain signals so that the computer may utilize them as input for a job. BCIs are interesting, but we must first comprehend the brain before diving into them.

BCIs are similar to computers, but what exactly is a computer? It is made up of transistors, hard drives, fans, displays, and other components. A computer is made up of several technologies that operate together. BCIs are similar, but they're maybe more complicated because they involve a combination of neurobiology and engineering. The knowledge is worthless when it is first acquired; it must be translated. Preprocessing is the term for this type of translation. We can extract the valuable information required for a task once the data has been preprocessed. Professors will offer enormous quantities of material just to test on a tiny portion of it, similar to how they will provide large amounts of information only to test on a small portion of it in university. The better pupils will concentrate on that tiny detail in order to prepare better. Feature extraction is the term for this procedure. Researchers may have sought to exploit a patient with locked-in syndrome's blinks as a form of communication. Data must be gathered and the essential information retrieved in order to do additional analysis. The machine learning element of BCIs is this additional analysis. BCIs (brain-computer interfaces) is a new technology that links the human brain to a computer or machine (BMIs) and interprets brain signals so that the computer may utilize them as input for a job.

From bionics to mind control, BCIs are changing the way people live, and they're growing more visible every day.

XVI

Questions

Consider the power of first impressions as you plan your outfit for your next debate. Why are they so impactful, and should they be? What does it take to change an impression? First impressions are extremely powerful in our ever-fleeting superficial world. Looks and the image you create can take you a long way on the road you want to traverse. This doesn't necessarily refer to physical appearance alone. Expressiveness through speech, posture and general presence too plays a vital role in making an impactful first impression.

They are judgments drawn at a split second and it takes quite some time to change an impression. If it has had a strong imprint on the other person, it might take a few years in my experience. You continually have to keep trying to make them somehow brush away the first impression. It is important to look confident and convincing during a debate to make your point reach the opponent. When you dress well people immediately perceive you to be educated and feel like you know what you are doing without you talking to them. It creates a memory and favorable memories have better outcomes in the future.

The order in which we encounter information has an impact on how we evaluate the following information. The halo effect, in which the perception of favourable characteristics in one object or part leads to the perception of comparable qualities in related things or in the whole, is connected to the exaggerated influence of first impressions.

You meet a nice individual at a party and are subsequently requested to raise funds for a good cause. You make touch with that individual because you believe she can help. There is no intrinsic link between being nice and being giving in reality. The halo effect, on the other hand, encourages you to believe that the two are connected. Most people believe that if she's good in one category (sociable), she'll be good in another (generous).

Consider the case of highly superior autobiographical memory, a condition in which people are unable to forget even the most mundane details of their daily lives. Is the perspective in this article too critical? Is it always better to have a better memory, or is it better to selectively (or unselectively) forget? It is better to have a better memory, even the mundane details as we never know if they can come to use in the future. Remembering every little thing allows you to pay attention to details and thus we start appreciating small things that take place and it is also useful to track down a particular memory when the whole sequence of memories are present almost exactly the way they actually took place. Forgetting has its own merits. For an individual it is better to selectively unconsciously forget aspects/ memories that are harmful to the individual's psyche. When memories never fade, the past can poison the present.

Having superior autobiographical memory does sound pretty appealing, doesn't it? How cool would it be if you could clear all your tests with just a glance at the textbook? But on the other hand, how gruesome would it be if you remember every moment of almost every day you've ever lived? That includes everything you've seen, heard and felt (both physically and emotionally). There are situations that we all would like to forget. Sometimes we hope we never saw/got to know about something in the first place (AND ALL THOSE SUPER EMBARRASSING MOMENTS YOU'RE DYING TO FORGET). In scenarios like that, having such a powerful memory would truly be a bane rather than a boon.

Music is extremely powerful as a summoner of memories because of its emotional connectivity. Can any other types of stimuli (perhaps based on other senses) also elicit such vivid memories? The smell of a person's perfume or their smell, in general, can trigger emotions and feelings that we thought to have been forgotten. Also, touch. A person's touch can also bring memories from the past and in the end, all memories are connected to emotions, we feel that emotion we experienced all over again, almost like deja vu. Any song that I know will automatically cause a reaction however some are linked to specific memories. Smells can be linked as well. If I smell mashed potatoes or broth soup I'm reminded of family. Every stimulus that's associated with our senses(taste, sight, hearing, and especially touch) has the capability to evoke certain nice and distasteful memories in us. The brain is just beautiful, to say the least.

XVII

Cases

The weapons effect is a social psychology phenomenon in which the mere sight of a weapon or an image of one causes individuals to become more violent. The initial research is both intriguing and ominous.

A 100 male university students were gathered and each was administered 1 shock or 7 based on random assigning. They were told that these shocks came from a peer. After that, the researchers let the participants give the peer as many shocks as they wanted. A revolver and rifle were in close proximity to a third of these pupils, and half of them were informed the weapons belonged to a classmate, while the other half were told they were merely there.

Two-thirds had no weapons on them or had two badminton rackets. Students who were in their presence administered more shocks, regardless of whether the weapons belonged to the peer or not. As a result, the two psychologists came to the conclusion that in the presence of firearms, and an aroused individual would act more violently.

In the scientific subject of social psychology, the phenomenon is described and shown. It relates to how the sheer sight of a weapon or a picture of a weapon causes individuals to become more violent, especially if they are already aroused.

ॐ

The car crash theory is intriguing because it demonstrates how a seemingly insignificant alteration may result in a faulty memory. It is made up of two experiments. The first task required 45 American students to watch seven short clips (5-30 seconds each) and explain what happened as though they were eyewitnesses to the disaster.

They were then given a series of questions, one of which was crucial: "About how fast were the automobiles driving when they (smashed/collided/bumped/hit/made contact)?" Notice how the phrasing was different, with these five verbs serving as the foundation for the experiment's conclusion.

The word "smashed" was found to have the highest speed estimate of 40.8mph, followed by "collided" with 39.3mph, "bumped" with 38.1mph, "struck" with 34.1mph, and lastly "made contact" with 31.8mph.

This demonstrated that the participants' perceptions may be influenced by a simple verb. A one-minute clip of another vehicle accident was presented to 150 students in the second experiment. Following that, 50 of them were asked 'how fast were the vehicles going when they collided?', another 50 were asked 'how fast were the cars driving when they smashed each other?', and the remaining 50 were not given any questions at all as the control group.

They answered 10 questions, one of which was a key one randomly inserted in the list: "Did you notice any shattered glass?" one week later the dependent variable was assessed.

"yes or no?" In the original film, there was no shattered glass. When asked how fast the automobiles were driving when they smashed, participants were more likely to say they saw broken glass. The real data table is at the bottom. The study's major finding was that interrogation tactics and information gathered after the incident might mix with the original memory, resulting in an incorrect recollection.

The Einstellung effect happens when one's capacity to achieve an ideal solution is hampered by prior information. When we believe we already know a solution, even if it isn't true or optimum, we become unable to explore alternative options. It renders us cognitively unable of distinguishing between prior and current problems. As a result, we may fix a problem but we do not innovate.

The brain tries to work effectively by referring to previous answers rather than focusing on the current challenge. It's trapped in a mentality. Instead of analyzing the problem on its own terms, we apply prior approaches to an apparently comparable situation. This impact may be seen in a variety of disciplines and at various skill levels. We all go through it, whether we realize it or not.

The water jar problem, undertaken by Abraham Luchins in 1942, is the classic experiment used to validate this effect. Participants were divided into two groups, one of which was given a few priming questions before being asked the main question. The priming questions drew the first group's attention to a certain approach to tackling the problem.

They were unable to tackle the fundamental problem, which could not be solved using the same approach as before. Participants in the second group, on the other hand, were presented with the identical core issue without the primer and were able to identify the best response more frequently than not.

Another investigation looked at the eye movements of chess players on the board. The participants were divided into two groups, one with a suboptimal answer on the board alongside an optimum solution, and the other with only the optimal solution on the board. Even though they stated that they were actively seeking a better solution, the group with the poor solutions continued to look at squares relevant to the found answer. Their gaze was drawn to the well-known answer. Even though they were attempting to evaluate the board objectively, the Einstellung effect prohibited them from doing so.

This impact implies that the more experience we gather, the more likely we are to succumb to its influence and fail to assess each situation on its own merits. We must determine what the underlying difference between this problem and the previous one is, and then analyze each new problem objectively. Prevent our brains from being automated and falling into autopilot mode. These mistakes are caused by initial beliefs developed from prior experience, not by a lack of information.

ଞ

A big, hefty middle-aged guy stole two Pittsburgh banks in broad daylight one day in 1995. He didn't hide behind a mask or any other disguise. Before leaving each bank, he grinned at the security cameras. Police nabbed a surprised McArthur Wheeler later that night. Wheeler was taken aback when the surveillance videos were shown to him. He muttered, 'But I wore the juice.' Wheeler apparently believed that putting lemon juice on his skin would make him undetectable to videotape cameras. After all, lemon juice is used as invisible ink, so as long as he didn't get too close to a heat source, he should have been undetectable.

Police concluded that Wheeler was not crazy or on drugs – just incredibly mistaken.

While virtually everyone has a positive opinion of their skills in different social and cognitive domains, some people incorrectly believe their abilities are far greater than they are. The 'Dunning-Kruger effect,' which explains the cognitive bias to exaggerate self-assessment, is now known as the 'illusion of confidence.' Dunning and Kruger (the scientists after whom the effect is named) devised several ingenious experiments to examine this phenomenon in the lab. They offered undergraduate students a series of questions on grammar, logic, and humor, and then asked each student to evaluate his or her total score, as well as their relative rank among the other students, in one research.

Surprisingly, pupils who performed poorly on these cognitive tests consistently underestimated their performance — by a large margin. Students in the poorest quartile thought they had done better than two-thirds of the class! Sure, it's common for people to exaggerate their ability. According to one research, 80% of drivers consider themselves above average, which is statistically impossible. When people estimate their relative popularity and cognitive ability, similar patterns have been seen.

The issue is that when individuals are incompetent, they not only come to incorrect conclusions and make poor decisions, but they also lose the ability to recognize their errors.

ಙ

Time capsules, such as the Crypt of Civilization, can allow the past to "communicate" with the future.

A stainless steel vault door was welded shut over sixty-five years ago in the basement of Phoebe Hearst Hall at Oglethorpe University in Georgia. Behind this door is a waterproofed chamber filled with a variety of once-modern objects and microfilm documents donated by men and women between 1937 and 1940. If they succeed, the vault's contents will remain hidden and undisturbed for the next 6,107 years The Crypt of Civilization is an ambitious project that began at the start of World War II and marks the first coordinated effort to gather and preserve a picture of human civilization and technology.

The initiative seemed to inspire and worry the people at the same time. The enormity of the concept and its far-reaching implications were unprecedented in contemporary history. The crypt has been featured in several newspapers and radio programs throughout the world. Because poison darts and gigantic stone spheres were not in the budget, the architects of this crypt utilized a new deterrent against would-be early intruders: guilt. A stainless steel sign hangs above the crypt's locked door, imploring everyone who enters to respect the crypt's contents until the year 8113.

The plaque reads:

This Crypt contains memorials of the civilization which existed in the United States and the world at large during the first half of the twentieth century. In receptacles of stainless steel, in which the air has been replaced by inert gasses, are encyclopaedias, histories, scientific works, special editions of newspapers, travelogues, travel talks, cinema reels, models, phonograph records, and similar materials from which an idea of the state and nature of the civilization which existed from 1900 to 1950 can be ascertained.

No jewels or precious metals are included. We depend upon the laws of the county of DeKalb, the State of Georgia, and the government of the United States and their heirs, assigns, and successors, and upon the sense of sportsmanship of posterity for the continued preservation of this vault until the year 8113, at which time we direct that it shall be opened by authorities representing the above governmental agencies and the administration of Oglethorpe University. Until that time we beg of all persons that this door and the contents of the crypt within may remain inviolate.

Of course, if an archaeologist today discovered a buried

tomb with an inscription instructing the finder not to open the entrance until a later period, the tomb's creators' intentions are unlikely to be honored. Given the Crypt of Civilization's waterproofed subterranean position and surrounding granite terrain, the vault is likely to survive until the year 8113 A.D. The area does not experience much seismic activity, and when earthquakes do occur, they are neither as often or as severe as they are in many other places of the world. The state of the crypt's contents is unknown at this time, although the artifacts inside are thought to be in good shape. However, the 65 years since its door was welded shut are a blip on the radar compared to the 6,107 years it has still to go.

BIBLIOGRAPHY

[1]https://avansalpacaresources.weebly.com/ science-2018.html

[2]https://markw1965.wordpress.com/world-scholars-cup-2018/science-the-science-of-memory/

[3]https://www.scholarscup.org/subjects/2018/science/

[4]https://explorable.com/ priming#:~:text=Priming%20is%20the%20implicit %20memory,in%20positive%20and%20negative%20ways.

[5]https://www.simplypsychology.org/multi-store.html

[6]]https://www.helpguide.org/articles/anxiety/panic-attacks-and-panic-disorders.htm

[7]https://my.clevelandclinic.org/health/diseases/ 12119-insomnia

[8]https://www.psychestudy.com/cognitive/memory/ motivated-forgetting

[9]https://www.webmd.com/schizophrenia/guide/ what-is-psychosis

[10] https://www.mayoclinic.org

[11] https://www.alz.org/alzheimers-dementia/what-is-dementia

[12]https://artofmemory.com/blog/simonides-of-ceos/

[13]https://www.healthline.com/nutrition/nootropics

[14]https://www.computerhope.com/jargon/p/ punccard.htm

[15]https://whatis.techtarget.com/definition/crawler

[16]https://developer.mozilla.org/en-US/docs/Web/ HTTP/Caching

[17] https://www.darpa.mil/program/restoring-active-memory

[18]https://www.technologyreview.com/2008/12/11/
95892/new-ways-to-boost-memory/

[19]https://futurism.com/scientists-use-light-alter-
memories-mice

[20]https://www.smithsonianmag.com/innovation/
meet-two-scientists-who-implanted-false-memory-
mouse-180953045/

[21]   https://www.nature.com/articles/
s41598-021-85134-4

[22]https://www.psychologytoday.com/intl/blog/am-i-
right/201302/the-power-first-impressions

[23]   https://www.damninteresting.com/the-crypt-of-
civilization/

[24]https://www.all-about-psychology.com/the-illusion-
of-competence.html

[25]https://www.exaptive.com/blog/einstellung-effect-0

[26]https://medium.com/geekculture/an-introduction-
to-brain-computer-interfaces-3c2b39873c5e

[27]]https://www.medicalnewstoday.com/articles/
249347#what-is-a-phobia

[28]https://www.betterhealth.vic.gov.au/health/
conditionsandtreatments/paranoia#what-is-paranoia

[29]https://www.apa.org/topics/anxiety

[30]https://rarediseases.info.nih.gov/diseases/6919/
locked-in-syndrome